Detectives:
Search for the Facts...

Helping Out

Anne O'Daly

Published by Brown Bear Books Ltd
4877 N. Circulo Bujia
Tucson, AZ 85718
USA

and

G14, Regent Studios
1 Thane Villas
London N7 7PH
UK

© 2025 Brown Bear Books Ltd

ISBN 978-1-83572-024-0 (ALB)
ISBN 978-1-83572-030-1 (paperback)
ISBN 978-1-83572-036-3 (ebook)

All rights reserved. No part of this book may be reproduced, stored in a retrieval system, or transmitted, in any form or by any means, electronic, mechanical, photocopying, recording, or otherwise, without the prior written permission of the copyright holder.

Library of Congress Cataloging-in-Publication Data available on request

Design Manager: Keith Davis
Children's Publisher: Anne O'Daly
Picture Manager: Sophie Mortimer

Picture Credits
Cover: Shutterstock: Carmen RL. Interior: Alamy: Stephen J. Kazlowski 8; Dreamstime: Outdoorsman 4b; iStock: Claudio Arnese 5tl, Artrush 14-15, Ken Canning 4-5c, 16-17, Jacob Eukman 9, Lavin Photography 20-21, skillpad 8-9, 22tl; Shutterstock: Andaman 12-13, Azrael3141 10-11, cbpix 4-5b, cpaulfell 5b, Divelvanov 5c, Angelo Saulo Ferreira 5tr, Judith Goswell 15, Ken Griffiths 6-7, Samy Kassem 13, Iaus Mohr 6, polarmaster 19, usey 16, Tara N Salgado 22bl, sasimoto 8br, SergeUWPhoto 1, 10, Wolfgang Simlinger 20, Adilson Sochodolak 22br, Liz Weber 21, Richard Whitcombe 22tr. Sheron Zhu 4t, 23.
t=top, b=bottom, l=left, r=right, c=center
All artwork and other photography Brown Bear Books.

Brown Bear Books has made every attempt to contact the copyright holder.
If you have any information about omissions, please contact: licensing@brownbearbooks.co.uk

Manufactured in the United States of America
CPSIA compliance information: Batch#AG/5663

Websites
The website addresses in this book were valid at the time of going to press. However, it is possible that contents or addresses may change following publication of this book. No responsibility for any such changes can be accepted by the author or the publisher. Readers should be supervised when they access the Internet.

Contents

Meet the Teams 4

Fact Files

    Leafcutter Ant 6

    Acacia Tree and Ants 8

    Cleaner Wrasse 10

    Clownfish and Sea Anemone 12

    Ostrich and Zebra 14

    Buffalo and Oxpecker 16

    Polar Bear and Arctic Fox 18

    Screech Owl and Blind Snake 20

Quiz 22

Useful Words 23

Find Out More 24

Index 24

Meet the Teams

Some animals work with others.
The animals can be very different.
But they help each other out.
Meet some amazing animal teams!

Screech Owl and Blind Snake

Buffalo and Oxpecker

Polar Bear and Arctic Fox

Ostrich and Zebra

Helping Out

Leafcutter Ant

Cleaner Wrasse

A cleaner wrasse cleans inside a bigger fish's mouth.

Clownfish and Sea Anemone

Acacia Tree and Ants

5

Leafcutter Ant

Leafcutter ants grow fungus in their nests. They feed the fungus. They keep it safe from pests. Young ants eat the fungus.

FACT FILE

Name: leafcutter ant

Where it lives: Central and South America

What it eats: adults feed on leaf sap; the young eat fungus

Type of animal: insect

Size: up to 0.75 inches (16 mm) long

Body: sharp jaws to cut leaves

The ants snip small bits off leaves

They carry them back to the nest

These ants are at the opening of their nest.

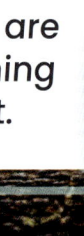

Helping Out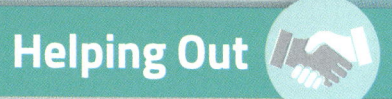

MINI FACTS

The ants can carry 50 times their body weight.

The ants feed the leaves to fungus

Acacia Tree and Ants

Acacia trees have sharp thorns. The thorns stop animals eating the leaves. The trees have another guard. Ants live inside the thorns.

FACT FILE

Name: acacia ant

Where it lives: on acacia trees in grasslands in Africa

What it eats: nectar from the acacia tree

Type of animal: insect

Size: 0.09 to 0.14 inches (2.5 to 3.5 mm)

Body: dark brown body, jaws can give a painful bite

The ants keep the tree safe

Giraffes eat acacia leaves. Their long tongues get around the thorns.

Helping Out

They get a safe place to live in return

The ants attack insects that try to eat the leaves

MINI FACTS

The acacia tree makes sugary nectar. The ants eat the nectar.

Sharp thorn

Cleaner Wrasse

These tiny fish live on coral reefs. They eat dead skin and bugs from bigger fish. The small fish get a meal. The big fish get cleaned.

FACT FILE

Name: cleaner wrasse

Where it lives: coral reefs in warm oceans

What it eats: dead skin from bigger fish

Type of animal: fish

Size: most cleaner wrasse are up to 7.9 inches (20 cm) long

Body: bright colors

The big fish opens its mouth so the cleaner wrasse can go inside

A cleaner wrasse nibbles near a fish's eye.

10

Helping Out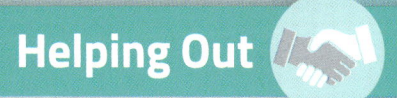

MINI FACTS

Some fish pretend to be cleaner wrasse. But they don't clean fish. They bite them instead!

Clownfish and Sea Anemone

Sea anemones have stingers. The stingers keep predators away. Clownfish live with sea anemones. The clownfish are safe from predators.

FACT FILE

Name: clownfish
Size: up to 6.5 inches (17 cm) long
Type of animal: fish
Name: sea anemone
Size: up to 3.9 inches (10 cm) long
Type of animal: invertebrate
Where they live: coral reefs in warm oceans

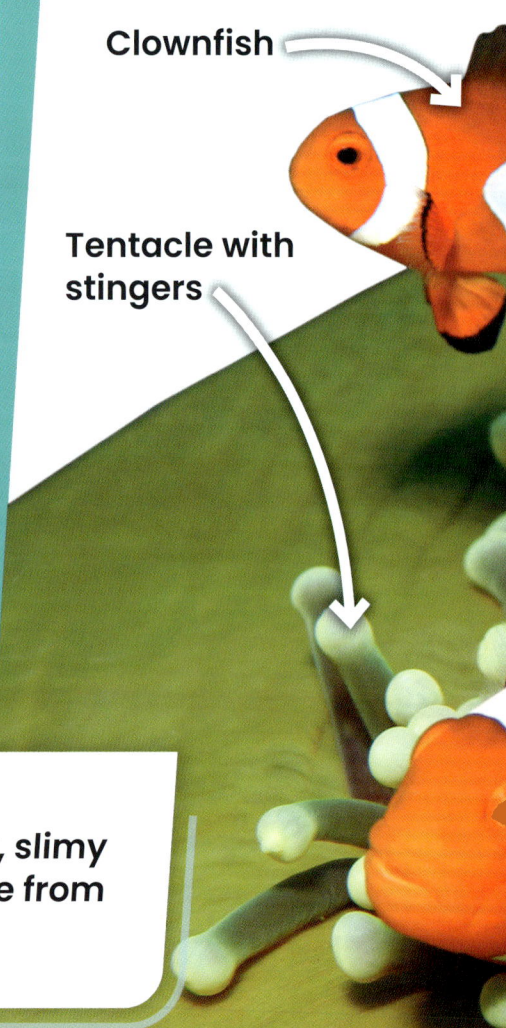

Clownfish

Tentacle with stingers

MINI FACTS

The clownfish has a thick, slimy coating. That keeps it safe from the anemone's stingers.

Helping Out

These crabs live with anemones, too. They help keep the anemones clean.

The sea anemone eats the clownfish's leftover food

Ostrich and Zebra

Ostriches and zebras live together.
Zebras have great hearing.
Ostriches have good eyesight.
They work as a team to look for danger.

An ostrich has big eyes and a long neck

FACT FILE

Name: ostrich

Size: up to 9 feet (2.7 m) tall

Type of animal: bird

Name: zebra

Size: 4 to 5 feet (1.2 to 1.5 m) tall at the shoulder

Type of animal: mammal

Where they live: grasslands in Africa

MINI FACTS

Ostriches don't have teeth.
They swallow stones.
The stones break up their food.

Helping Out

An ostrich can run at up to 43 mph (69 km/h).

Zebra ears move in different directions

Buffalo and Oxpecker

Oxpeckers sit on a buffalo's back.
They eat insects from the buffalo's skin.
The birds warn the buffalo if danger is near.

FACT FILE

Name: oxpecker

Size: 8 inches (20 cm)

Type of animal: bird

Name: African buffalo

Size: up to 5.6 feet (1.7 m) tall

Type of animal: mammal

Where they live: grasslands in Africa

Oxpeckers clean rhinos as well.

MINI FACTS

Buffalo live in herds. A herd can have up to 500 animals in it.

Helping Out

The oxpecker gets food

The buffalo is cleaned

17

Polar Bear and Arctic Fox

Polar bears are deadly hunters. They mainly eat fatty meat. Arctic foxes follow the bears. They eat any leftovers.

FACT FILE

Name: Arctic fox

Size: 18 to 27 inches (46 to 68 cm)

Type of animal: mammal

Name: polar bear

Size: up to 8 feet (2.4 m)

Type of animal: mammal

Where they live: the Arctic

The fox eats the meat left behind

MINI FACTS

Arctic foxes have babies in May and June. There can be 19 pups in a litter.

18

Helping Out

Ringed seals are a polar bear's favorite food.

Polar bears eat fatty animals for energy

19

Screech Owl and Blind Snake

Screech owls catch food for their babies. They take it to their nest. The owls take blind snakes to the nest, too. The snakes have a job to do!

FACT FILE

Name: Eastern screech owl

Size: up to 10 inches (25 cm)

Type of animal: bird

Name: Texas blind snake

Size: up to 3.9 inches (10 cm) long

Type of animal: reptile

Where they live: wooded places in North America

Blind snakes have smooth, slippery skin.

20

Helping Out

MINI FACTS

Screech owls hunt at night. They have good eyesight to see in the dark.

The snakes eat insect larvae in the nest

The snakes get a meal. The owl gets a clean nest.

Quiz

Test your skills!
Can you answer these questions?
The answers are on page 24.

① Why do acacia trees have sharp thorns?

② What keeps clownfish safe from a sea anemone's stings?

③ How do ostriches break up their food?

④ How do blind snakes help screech owls?

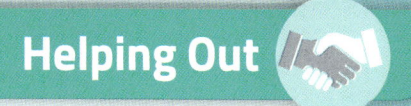

Helping Out

Useful Words

coral reef an underwater structure that is home to lots of animals

fungus a type of living thing that gets energy from rotting materials

herd a group of animals that eat and travel together

invertebrate an animal without a backbone

larvae (singular: larva) the young of an insect

litter baby animals that are born together

nectar a sugary liquid made by plants

predator an animal that hunts other animals for food

prey an animal that is hunted and eaten by other animals

tentacles bendy body parts

Find Out More

Books

Animal Teams: How Amazing Animals Work Together in the Wild, Charlotte Milner (DK Children, 2022)

Polar Bears and Arctic Foxes Team Up!, Stephanie Peters (Capstone Press, 2023)

Working as a Team, Nancy Dickmann (Brown Bear Books, 2021)

Websites

kids.kiddle.co/Cleaner_fish

kids.nationalgeographic.com/animals/fish/facts/clown-anemonefish

sdzwildlifeexplorers.org/animals/leafcutter-ants

Index

cleaning 5, 10–11, 16–17, 21

coral reefs 10, 12

fungus 6–7

getting food 13, 17–18, 21

looking out for danger 14–16,

nectar 8–9

safe place to live 9, 12

Quiz answers 1. Acacia trees have sharp thorns to stop animals eating the leaves; **2.** A clownfish has a thick, slimy coating that stops the stings from harming it; **3.** They swallow stones to break up their food **4.** They eat insect larvae in the nest.